CONTEMPORARY'S

Decimals

MATH EXERCISES

■ CONTENTS

Mc Graw Hill **Wright Group**

ISBN 0-8092-3826-8

Send all inquiries to:
Wright Group/McGraw-Hill
130 East Randolph, Suite 400
Chicago, Illinois 60601

Manufactured in the United States of America.

34 35 36 37 38 39 40 QDB 15 14 13 12

The McGraw-Hill Companies

■ INTRODUCING DECIMAL FRACTIONS

Dollars and Cents
Decimal fractions, like proper fractions, represent part of a whole. We use decimal fractions every day. This is because our money system is based on them.
- With dollars and cents, we use two decimal places:

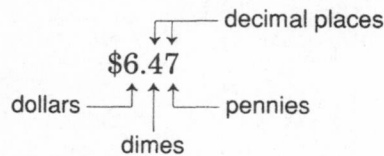

one-tenth of a dollar (.1)

one-hundredth of a dollar (.01)

- One dime ($0.10) is **one-tenth** of a dollar.
- One penny ($0.01) is **one-hundredth** of a dollar.

The First Three Decimal Places
From dollars and cents you already know the first two decimal place values: **tenths** and **hundredths**. The third decimal place value is **thousandths**.

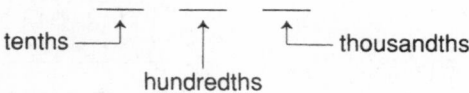

Examples:

0.3	3-tenths
0.7	7-tenths
0.04	4-hundredths
0.09	9-hundredths
0.006	6-thousandths
0.007	7-thousandths

The shaded part of each figure below is shown written as a decimal fraction.
A **leading zero** is written to the left of the decimal point to show no whole number.

1 part out of 10 parts — 0.1 (one-tenth) — leading zero

1 part out of 100 parts — 0.01 (one-hundredth)

1 part out of 1,000 parts — 0.001 (one-thousandth)

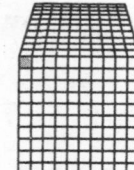

▶ **Write a decimal fraction to represent each of the following.**

1. 2 parts out of 10 parts _____

2. 6 parts out of 100 parts _____

3. 9 parts out of 1,000 parts _____

4. 5 parts out of 10 parts _____

5. 7 parts out of 100 parts _____

6. 3 parts out of 1,000 parts _____

Answers begin on page 30.

■ READING DECIMAL FRACTIONS

To read a decimal fraction:
• Read the number to the right of the decimal point. This number may have one or more digits.

• Read the place value of the digit at the far right.

Example	Number	+	Place Value	Read as
0.5	5		tenths	5-tenths
0.06	6		hundredths	6-hundredths
0.37	37		hundredths	37-hundredths
0.008	8		thousandths	8-thousandths
0.039	39		thousandths	39-thousandths
0.425	425		thousandths	425-thousandths

Although zero has no value, it is used as a **placeholder.**
• Placed between the decimal point and a digit, zero changes the value of a decimal fraction.

0.8 (eight-tenths) is not the same as 0.08 (eight-hundredths)

• Placed at the far right of a decimal fraction, a zero changes the way a fraction is read but does not change its value.

0.8 (eight-tenths) and 0.80 (eighty-hundredths) have the same value

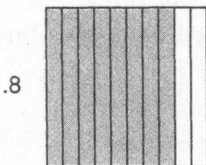

.8

has the same value as

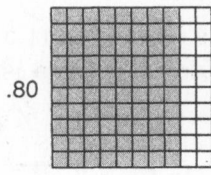

.80

▶ **Write the following decimal fractions in word form.**

1. 0.4 _____   0.8 _____

2. 0.07 _____   0.14 _____

3. 0.30 _____   0.50 _____

4. 0.062 _____   0.023 _____

5. 0.070 _____   0.090 _____

6. 0.146 _____   0.278 _____

7. 0.350 _____   0.600 _____

▶ **In each group, circle the two decimal fractions that have the same value.**

8. 0.4, 0.04, 0.040   9. 0.120, 0.012, 0.12   10. 0.70, 0.7, 0.07

Answers begin on page 30.

■ WRITING DECIMAL FRACTIONS

To write a decimal fraction:
• First, identify the place value of the right-hand digit.
• Then, write the number so that the right-hand digit is in its proper place, writing zeros as placeholders if necessary.

Example 1. Write *35-hundredths* as a decimal fraction.

Write 35 so that the 5 ends up in the hundredths place—the second place to the right of the decimal.

┌ Write a leading zero.
↓
0.35
   └ Write 5 in the hundredths place.

Example 2. Write *92-thousandths* as a decimal fraction.

Write 92 so that the 2 ends up in the thousandths place—the third place to the right of the decimal.

┌ Write zero as a placeholder.
↓
0.092
   └ Write 2 in the thousandths place.

▶ **Write each of the following as a decimal fraction.**

1. 5-tenths _____   7-tenths _____   9-tenths _____

2. 8-hundredths _____   9-hundredths _____   12-hundredths _____

3. 10-hundredths _____   40-hundredths _____   90-hundredths _____

4. 5-thousandths _____   23-thousandths _____   80-thousandths _____

5. 275-thousandths _____   750-thousandths _____   500-thousandths _____

6. thirty-five hundredths _____   one hundred ninety thousandths _____

7. eight-tenths _____   sixty-hundredths _____

8. three hundred forty thousandths _____   five hundred two thousandths _____

9. four-hundredths _____   two hundred seven thousandths _____

▶ **Write each underlined amount as a decimal fraction.**

10. Thirty-eight cents is <u>thirty-eight hundredths</u> of a dollar.   _____

11. One kilometer is about <u>six-tenths</u> of a mile.   _____

12. One yard is about <u>ninety-one hundredths</u> of a meter.   _____

13. One centimeter is about <u>three hundred ninety-four thousandths</u> of an inch.   _____

14. One pound is about <u>forty-five hundredths</u> of a kilogram.   _____

Answers begin on page 30.

■ COMPARING AND ORDERING DECIMAL FRACTIONS

To compare decimal fractions, begin in the tenths place and compare each digit. If two decimal fractions do not have the same number of digits, give them an equal number by adding one or more place-holding zeros.

Use the following symbols when comparing numbers.

< means "is less than"	$5 < 9$
> means "is greater than"	$8 > 4$
= means "is equal to"	$7 = 5 + 2$

Compare 0.62 and 0.58.	Compare $0.53 and $0.59.	Compare 0.24 and 0.245.
		Add a 0.
0.6̲2 Because 6 > 5,	$0.5̲3 Because 3 < 9,	0.240̲ Because 0 < 5,
0.5̲8 0.62 > 0.58.	$0.5̲9 $0.53 < $0.59.	0.245̲ 0.24 < 0.245.
	same first digit	same first two digits

▶ **A. Write >, <, or = to compare each pair of decimal fractions below.**

1.   0.5 _____ 0.7   0.43 _____ 0.4   0.76 _____ 0.93

2.   0.6 _____ 0.594   0.17 _____ 0.208   0.168 _____ 0.65

3. $0.05 _____ $0.13 $0.27 _____ $0.27 $0.51 _____ $.19

▶ **B. List the brands of orange juice concentrate in order of price per ounce (oz).**

Brands	Price		Listed by Unit Price
OJ Plus:	$0.094 per oz.	1st: _____	(least expensive)
Sunripe:	$0.089 per oz.	2nd: _____	
Emerald:	$0.09 per oz.	3rd: _____	
Valley D:	$0.102 per oz.	4th: _____	(most expensive)

▶ **C. Indicate with a check (✔) the drawer in which each crystal sample should be placed according to weight. (Note: kg = kilogram)**

Crystal Sample	Weight (in kilograms)	Drawer 1 (less than 0.2 kg)	Drawer 2 (between 0.2 and 0.45 kg)	Drawer 3 (more than 0.45 kg)
A	0.49	_____	_____	_____
B	0.072	_____	_____	_____
C	0.503	_____	_____	_____
D	0.4	_____	_____	_____
E	0.18	_____	_____	_____

Answers begin on page 30.

◼ READING AND WRITING MIXED DECIMALS

A **mixed decimal** is a whole number plus a decimal fraction. To read a mixed decimal, read the whole number first, say "and," and then read the decimal fraction. You read the decimal point as the word "and" when saying a money amount or a number.

Read $6.58 as "Six dollars **and** fifty-eight cents."

Read 6.58 as "Six **and** fifty-eight hundredths."

Usually, mixed decimals and decimal fractions are simply called "decimals."

Comparing Mixed Decimals
• To compare mixed decimals, compare the whole numbers first.
a) $7.82 > $5.96 because 7 > 5.   b) 91.6 < 93.2 because 91 < 93.

• If the whole numbers are the same, compare the decimal fractions.
a) $4.68 < $4.73 because 6 < 7.   b) $2.13 > $2.08 because 1 > 0.

▶ **Write the following as mixed decimals.**

1. five and three-tenths _____   fifteen and thirty-two hundredths _____

2. six and nineteen-thousandths _____   three and eight-hundredths _____

3. seventy-four and two hundred twenty-five thousandths _____

4. one hundred twenty-six and seven hundred forty-six thousandths _____

▶ **Write the following mixed decimals in word form.**

5. 5.7 _____   6.15 _____

6. 64.375 _____

7. 128.92 _____

▶ 8. **List the runners of each race in order of finish.**
Remember: the shorter the time, the better the finish.

Race 1 Times	Race 1 Results	Race 2 Times	Race 2 Results
Alan: 58.234 sec.	1st: _____	Stacey: 44.65 sec.	1st: _____
Vic: 59.084 sec.	2nd: _____	Jani: 43.7 sec.	2nd: _____
Sal: 58.79 sec.	3rd: _____	Ellie: 44.732 sec.	3rd: _____
Don: 59.06 sec.	4th: _____	Jolene: 43.78 sec.	4th: _____

Answers begin on page 30.

■ ESTIMATING WITH MIXED DECIMALS

Estimation is useful anytime you don't need an exact answer. It also can be helpful when choosing an answer to a multiple-choice question on a test. Estimation also can help when you're not sure where to place the decimal point in an exact answer.

To estimate with mixed decimals, round each mixed decimal to the nearest whole number. Then do the indicated math.

To round a mixed decimal to a whole number, look at the digit in the tenths place.
• If the digit is 5 or more, round up to the next larger whole number.
• If the digit is less than 5, drop all decimal digits. The whole number stays the same.

┌ tenths place
16.8 rounds to 17
└ 5 or more

┌ tenths place
5.39 rounds to 5
└ less than 5

┌ tenths place
9.542 rounds to 10
└ 5 or more

16.8 ≈ 17*

5.39 ≈ 5*

9.542 ≈ 10*

* The symbol ≈ means "is approximately equal to."

Example 1. Estimate 8.42 + 9.8

Step 1. Round each mixed decimal to the nearest whole number.

8.42 ≈ 8   9.8 ≈ 10

Step 2. Add the whole numbers.

8 + 10 = 18

8.42 + 9.8 ≈ 18

Example 2. Estimate 5.03 × 3.89

Step 1. Round each mixed decimal to the nearest whole number.

5.03 ≈ 5   3.89 ≈ 4

Step 2. Multiply the whole numbers.

5 × 4 = 20

5.03 × 3.89 ≈ 20

▶ **Round each mixed number to the nearest whole number or dollar.**

1. 5.7 meters ≈ $8.25 ≈   9.5 centimeters ≈

2. $17.40 ≈   7.82 ounces ≈   26.4 miles per gallon ≈

▶ **Choose the best estimate for each problem.**

3. 13.56 + 9.8 =

(1) 12 + 10

(2) 13 + 10

(3) 14 + 10

4. 34.5 − 12.9 =

(1) 34 − 12

(2) 35 − 12

(3) 35 − 13

5. 14.45 × 9.13 =

(1) 14 × 9

(2) 15 × 9

(3) 15 × 10

6. 34.1 ÷ 5.9 =

(1) 34 ÷ 6

(2) 34 ÷ 5

(3) 35 ÷ 6

9

▶ Find an estimated answer for each problem.

Estimate | Estimate | Estimate

7. $9.64 ≈
+ 3.87 ≈

7.3
− 4.325

$10.27
× 16.5

Estimate | Estimate | Estimate

8. 12.89
× 0.97

3.1)15.125

12.14)36.75

▶ Use an estimate to decide where to place the decimal point in each answer.

9. $9.38 + 4.7 + 2.875 = 16955$

10. $24.625 − 12.94 = 11685$

11. $15.25 × 6.13 = 934825$

12. $39.4 × 10.3 = 40582$

13. $36.624 ÷ 4.8 = 763$

14. $124.293 ÷ 3.9 = 3187$

▶ Estimate an answer to each problem. Then use your estimate as a guide to help you choose the exact answer from the choices given.

15. Three packages are weighed on a metric scale. The three weights are 1.925 kg, 2.25 kg, and 6.8 kg. What is the total weight of these three packages?

(1) 8.755 kg
(2) 9.325 kg
(3) 10.975 kg
(4) 12.525 kg
(5) 14.250 kg

16. As a secretary, Jani earns $7.19 per hour. How much will Jani earn in a week in which she works 39.75 hours?

(1) $264.30
(2) $285.80
(3) $293.40
(4) $308.60
(5) $319.20

17. When beef roasts went on sale for $3.14 per pound, Jaclyn bought one that weighed 7.8 pounds. What total price should Jaclyn be charged?

(1) $16.39
(2) $18.99
(3) $20.89
(4) $22.19
(5) $24.49

18. Alan drove 159.6 miles on 7.9 gallons of gas. What was his car's approximate mileage (miles per gallon) on this drive?

(1) 20.2
(2) 22.3
(3) 24.6
(4) 26.5
(5) 28.8

Answers begin on page 30.

■ ROUNDING TO A CHOSEN PLACE VALUE

Smaller Place Values

Once in a while, you may see a decimal fraction with more than three decimal places. Because of this, you should be familiar with place values smaller than 0.001.

0.1 one-tenth
0.01 one-hundredth
0.001 one-thousandth
0.0001 one ten-thousandth
0.00001 one hundred-thousandth
0.000001 one-millionth

Decimal Place Values

```
0. ____  ____  ____    ____  ____  ____
tenths ⌐              millionths ⌐
  hundredths ⌐          hundred-thousandths ⌐
    thousandths ⌐    ten-thousandths ⌐
```

Rounding to a Chosen Place Value

As was discussed on the previous two pages, **rounding** mixed decimals *before doing the math* is a good way to estimate an answer.

Rounding also is used to simplify an exact answer *after the math is done*. Most often, an answer is rounded to a chosen place value, giving the answer fewer decimal digits.

To round a decimal fraction to a chosen place value, look at the digit to the **right** of the chosen place value.
• If the digit is *5 or more*, *round up*.
• If the digit is *less than 5*, *leave* the digit in the chosen place value *unchanged*.

Rounding to the tenths place (nearest tenth)

Check the digit in the hundredths place.

$0.65 \approx 0.7$ $2.61 \approx 2.6$ $8.672 \approx 8.7$

   ⌐ 5 or more   ⌐ less than 5   ⌐ 5 or more

Rounding to the hundredths place (nearest hundredth)

Check the digit in the thousandths place.

$0.429 \approx 0.43$ $2.384 \approx 2.38$ $4.168 \approx 4.17$

   ⌐ 5 or more   ⌐ less than 5   ⌐ 5 or more

Rounding to the thousandths place (nearest thousandth)

Check the digit in the ten-thousandths place.

$0.2756 \approx 0.276$ $1.0852 \approx 1.085$ $3.4635 \approx 3.464$

   ⌐ 5 or more   ⌐ less than 5   ⌐ 5 or more

▶ **Round each number to the tenths place.**

1. 0.38 ≈   0.29 ≈   0.53 ≈   0.147 ≈   0.824 ≈

2. 3.25 ≈   1.93 ≈   2.804 ≈   3.062 ≈   7.952 ≈

▶ **Round each number to the hundredths place.**

3. 0.235 ≈   0.812 ≈   0.349 ≈   0.892 ≈   0.246 ≈

4. 2.825 ≈   1.052 ≈   6.500 ≈   7.806 ≈   21.875 ≈

▶ **Round each number to the thousandths place.**

5. 0.4737 ≈   0.9734 ≈   1.9472 ≈   3.4889 ≈   17.3375 ≈

6. 0.3075 ≈   0.6403 ≈   4.2507 ≈   6.0089 ≈   12.1052 ≈

7. A grocery clerk used a calculator to multiply *pounds (lb.)* by *dollars per pound ($ per lb.)* to find the selling price of each package of beef listed on the chart below. Her exact calculator answers are shown.

Find the selling price of the packages by rounding each calculator answer to the nearest cent (hundredths place).

	Weight (lb.)		Price ($ per lb.)	Calculator Answer	Selling Price
Hamburger	4.05	×	$1.83	7.4115	(a) $_____
Round steak	7.1	×	$2.19	15.549	(b) _____
Sirloin steak	3.90	×	$3.88	15.132	(c) _____
Rib steak	2.75	×	$4.09	11.2475	(d) _____
T-bone steak	2.1	×	$5.12	6.894	(e) _____
Tenderloin steak	2.08	×	$6.15	12.792	(f) _____
Note: A calculator does not display a dollar sign ($).					

(g) The clerk made a mistake when writing one of the calculator answers. Using estimation, determine which calculator answer is incorrect.

Answers begin on page 31.

■ ADDING DECIMALS

To add decimals:
• Line up the decimal points and add the columns from right to left.
• Place a decimal point in the answer directly below the other decimal points.

Example 1. 0.64 + 0.39

Line up the   Add the columns.
decimal points.

  ↓   1 1
 0.64   0.64
+ 0.39   + 0.39
             1.03
              ↑— Place a decimal
                 point in the answer.

Example 2. 5.74 + 3.38

Line up the   Add the columns.
decimal points.

  ↓   1 1
 5.74   5.74
+ 3.38   + 3.38
             9.12
              ↑— Place a decimal
                 point in the answer.

▶ **Add. Use estimating to check problems involving mixed decimals.**

1.
0.9	0.8	0.46	1.86	0.746	2.395
+ 0.6	+ 0.7	+ 0.39	+ 0.75	+ 0.585	+ 0.608

2.
6.3	17.9	24.75	35.85	7.808	14.581
+ 2.8	+ 9.6	+ 11.62	+ 20.47	+ 5.261	+ 8.627

3.
$0.75	$0.90	$2.58	$3.45	$7.50	$25.00
+ 0.35	+ 0.56	+ 0.49	+ 1.75	+ 3.28	+ 9.64

4.
2.4	9.5	$12.75	$34.86	4.382	12.458
0.7	6.8	5.67	23.90	2.472	8.250
+ 0.6	+ 0.9	+ 0.48	+ 12.65	+ 1.750	+ 4.837

▶ **Line up the decimal points and add. Round the answers to row 5 to the tenths place.**

5. 0.85 + 0.46 =   3.43 + 0.94 =   7.85 + 4.62 =   23.45 + 9.54 =

6. $1.50 + $0.90 = $45.64 + $32.86 = $7.48 + $5.46 =

Answers begin on page 31.

■ USING ZEROS AS PLACEHOLDERS

To add decimals that do not have the same number of decimal places, use zeros as **placeholders**. Remember: a whole number is "understood" to have a decimal point to the right of the ones digit.

Example 1. 3.2 + 1.75

Line up the decimal points.

$$\begin{array}{r} 3.2 \\ + 1.75 \end{array}$$

Write a place-holding zero in the top number. Add.

$$\begin{array}{r} 3.20 \\ + 1.75 \\ \hline 4.95 \end{array}$$

└── Place a decimal point in the answer.

Example 2. 4 + 3.625

Write 4 as 4. and line up the decimal points.

$$\begin{array}{r} 4. \\ + 3.625 \end{array}$$

Write three place-holding zeros in the top number. Add.

$$\begin{array}{r} 4.000 \\ + 3.625 \\ \hline 7.625 \end{array}$$

└── Place a decimal point in the answer.

▶ **Add. Use estimating to check problems involving mixed decimals.**

1.
0.55	0.75	0.8	0.9	0.375	0.865
+ 0.4	+ 0.9	+ 0.27	+ 0.835	+ 0.25	+ 0.5

2.
4	5	6.5	9.8	5	9.667
+ 3.2	+ 4.1	+ 6	+ 3	+ 3.33	+ 7

3.
6.46	7.2	8.625	4.9	9.82	23.5
+ 2.1	+ 3.54	+ 5.5	+ 1.325	+ 2.4	+ 9.64

4.
5.25	9.5	8	12.4	4	15.4
0.7	6	5.67	11	2.4	8.253
+ 0.6	+ 3.25	+ 2.4	+ 8.65	+ 1.75	+ 4.83

▶ **Line up the decimal points and add. Round the answers in row 5 to the hundredths place.**

5. 7.625 + 4.5 + 2.35 =   8 + 3.875 + 1.5 =   6.5 + 5 + 3.009 =

6. $4 + $0.75 = $8.25 + $6 = $12 + $10.48 + $5.38 =

■ SUBTRACTING DECIMALS

To subtract decimals:
- Write the smaller number below the larger number and line up the decimal points.
- Subtract the columns from right to left, regrouping if necessary.
- Place a decimal point in the answer directly below the other decimal points.

Example 1. 0.86 − 0.49

Line up the
decimal points.

$$\begin{array}{r} \downarrow \\ 0.86 \\ -\ 0.49 \\ \hline \end{array}$$

Regroup the tenths.
Subtract the columns.

$$\begin{array}{r} 7\ 16 \\ 0.8\,6 \\ -\ 0.4\,9 \\ \hline 0.3\,7 \end{array}$$

└── Place a decimal
point in the answer.

Example 2. 6.4 − 3.8

Line up the
decimal points.

$$\begin{array}{r} \downarrow \\ 6.4 \\ -\ 3.8 \\ \hline \end{array}$$

Regroup the ones.
Subtract the columns.

$$\begin{array}{r} 5\ 14 \\ 6.4 \\ -\ 3.8 \\ \hline 2.6 \end{array}$$

└── Place a decimal
point in the answer.

▶ **Subtract. Use estimating to check problems involving mixed decimals.**

1.

0.9	0.7	0.87	0.76	0.345	0.750
− 0.5	− 0.2	− 0.25	− 0.38	− 0.127	− 0.255

2.

8.4	9.3	9.8	12.67	17.52	24.50
− 5.3	− 4.1	− 7.9	− 8.45	− 9.67	− 16.75

3.

9.42	8.36	5.00	14.576	8.000	24.00
− 2.84	− 6.79	− 2.75	− 5.625	− 3.382	− 16.75

4.

$0.88	$0.90	$1.68	$3.54	$8.30	$15.00
− 0.26	− 0.48	− 0.79	− 1.65	− 4.28	− 6.89

▶ **Line up the decimal points and subtract.**

5. $0.9 − 0.2 =$ $7.6 − 0.8 =$ $9.3 − 2.7 =$ $17.37 − 9.59 =$

6. $\$3.70 − \$0.85 =$ $\$23.80 − \$11.57 =$ $\$35.00 − \$15.26 =$

To subtract decimals that do not have the same number of decimal places, write in place-holding zeros. Use these zeros in **regrouping** in the same way you do when working with whole numbers.

Example 3.   4.79 − 2.5

Line up the decimal points.

Write a place-holding zero in the bottom number. Subtract.

$$\begin{array}{r} \downarrow \\ 4.79 \\ -\ 2.5 \\ \hline \end{array}$$

$$\begin{array}{r} 4.79 \\ -\ 2.50 \\ \hline 2.29 \end{array}$$

└── Place a decimal point in the answer.

Example 4.   9 − 5.75

Write 9 as 9. and line up the decimal points.

Write two place-holding zeros in the top number. Regroup and subtract.

$$\begin{array}{r} \downarrow \\ 9. \\ -\ 5.75 \\ \hline \end{array}$$

$$\begin{array}{r} {\scriptstyle 8\ 9\ 10} \\ 9.\!\!\not{0}\,\not{0} \\ -\ 5.75 \\ \hline 3.25 \end{array}$$

└── Place a decimal point in the answer.

▶ **Subtract. Use estimating to check problems involving mixed decimals.**

7.

0.75	0.84	0.375	0.945	0.654	0.825
− 0.6	− 0.3	− 0.25	− 0.6	− 0.3	− 0.5

8.

0.9	0.7	0.8	0.6	0.8	0.9
− 0.45	− 0.52	− 0.67	− 0.325	− 0.225	− 0.345

9.

8.8	6	5	13	9	29
− 3	− 4.5	− 2.75	− 4.625	− 6.735	− 14.25

▶ **Line up the decimal points and subtract.**

10. 0.85 − 0.6 =   0.675 − 0.4 =   0.9 − 0.75 =   0.8 − 0.325 =

11. 1.47 − 0.5 =   2.07 − 0.9 =   6 − 1.8 =   12.5 − 7.875 =

12. $5 − $0.75 = $10.00 − $6.43 = $25.00 − $13.79 =

Answers begin on page 31.

■ APPLYING ADDITION AND SUBTRACTION SKILLS

▶ **Use addition or subtraction to solve each problem.**

1. If unleaded gas is selling for $1.284 per gallon, how much would you pay for two gallons? Be sure to round your answer to the nearest cent. (In $1.284, the 4 stands for 4-tenths of a cent.)

2. During the first two months of the baseball season, Manny's batting average was .235. By the end of the season his average had increased by .074. What was Manny's batting average at the end of the season?

3. Rebecca's new skis are 1.67 meters long, 0.057 meter shorter than the skis she used last year. How long were the skis Rebecca used last year?

4. Each morning, Amelia drives 3.8 miles to drop her daughter off at school. She then drives another 2.75 miles to get to work. To the nearest tenth of a mile, how far does Amelia drive between home and work each morning?

5. At the Clothes Tree, Phil sells sweaters for $39.85, making a profit of $14.89 on each sweater. What does Phil actually pay for each sweater he sells?

6. A $\frac{1}{16}$-inch drill bit has a diameter of 0.0625 inch. The next size, a $\frac{5}{64}$-inch bit, has a diameter of 0.078125 inch. Rounding your answer to the nearest thousandth of an inch, determine how much larger in diameter the larger bit is than the smaller one.

7. A machinist needs to cut 0.625 inch off the radius of a shaft. After he has cut off 0.5 inch, how much more does he still need to cut?

8. Normal human body temperature is 98.6°F. When she had a fever, Shana's temperature reached 103°F. By how much had Shana's temperature increased at this point?

9. Ephran's dinner bill was $6.49. How much change should Ephran receive if he pays with a $10 bill?

10. Ben worked 7.25 hours on Monday, 7.875 hours on Tuesday and Wednesday, 8 hours on Thursday, and 8.5 hours on Friday. To the nearest hour, how much did Ben work during this week?

11. The Frankels pay $0.0836 for each kilowatt-hour of electricity they use. In the neighboring state, electric power costs $0.115 per kilowatt-hour. To the nearest cent, how much less do the Frankels pay per kilowatt-hour than people in the next state?

12. Josh's pickup is 5.914 meters long and his trailer is 4.12 meters long. Which of the following is the **best estimate** of the combined length in meters of the pickup and trailer?

    (1) 5 + 4
    (2) 6 + 5
    (3) 6 + 4
    (4) 5 + 6
    (5) 6 + 6

Answers begin on page 31.

■ MULTIPLYING DECIMALS BY WHOLE NUMBERS

To multiply a decimal by a whole number:
• Multiply the numbers just as you would whole numbers.
• Count the number of decimal places in the decimal to see how many decimal places are in the answer.
• Count off the number of decimal places, starting at the right of the answer, and then write the decimal point.

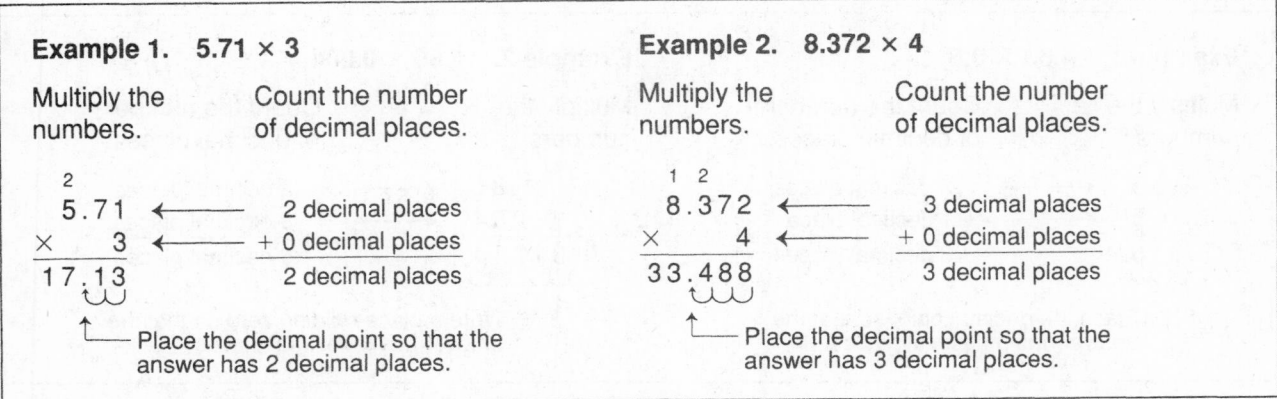

Example 1. 5.71 × 3

Multiply the numbers.   Count the number of decimal places.

```
    2
  5.71  ←————— 2 decimal places
×    3  ←————— + 0 decimal places
 17.13         2 decimal places
   ↺↺
     └— Place the decimal point so that the
         answer has 2 decimal places.
```

Example 2. 8.372 × 4

Multiply the numbers.   Count the number of decimal places.

```
   1 2
 8.372  ←————— 3 decimal places
×    4  ←————— + 0 decimal places
33.488         3 decimal places
   ↺↺↺
      └— Place the decimal point so that the
          answer has 3 decimal places.
```

▶ **Multiply. Use estimating to check problems involving mixed decimals.**

1.
```
  0.8        0.9        0.6        0.4        0.7        0.8
× 5        × 4        × 7        × 8        × 9        × 7
```

2.
```
  6.5        8.4        4.7        5.8        7.8        9.5
× 3        × 6        × 8        × 9        × 6        × 4
```

3.
```
  3.42       7.54       8.05       9.54       7.425      8.737
×   8      ×   2      ×  14      ×  12      ×   15     ×   21
```

4.
```
$0.80      $0.75      $ 0.48     $2.75      $8.50      $15.00
×   5      ×   6      ×   9      ×  17      ×  28      ×   14
```

▶ **Multiply. Round each answer to the tenths place.**

5. 0.91 × 8 =   8.32 × 7 =   3.26 × 12 =   8.625 × 15 =

Answers begin on page 31.

18

■ MULTIPLYING DECIMALS BY DECIMALS

To multiply a decimal by a decimal:
- Multiply the numbers just as you would whole numbers.
- Count the number of decimal places in **both decimals** to see how many decimal places are in the answer.
- Count off the number of decimal places, starting at the right of the answer.
- Write the decimal point, using place-holding zeros if necessary.

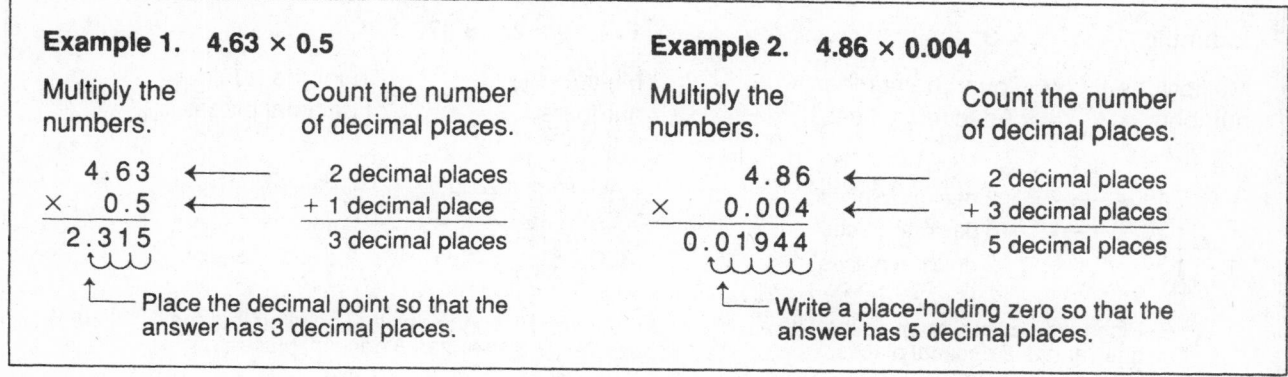

▶ **Multiply. Use estimating to check problems involving mixed decimals.**

1.
$$0.6 \times 0.4 \qquad 0.7 \times 0.2 \qquad 0.8 \times 0.5 \qquad 0.9 \times 0.06 \qquad 0.5 \times 0.02 \qquad 0.6 \times 0.006$$

2.
$$7.5 \times 0.4 \qquad 9.3 \times 0.6 \qquad 5.9 \times 0.7 \qquad 6.8 \times 0.08 \qquad 5.4 \times 0.05 \qquad 8.5 \times 0.007$$

3.
$$3.42 \times 0.05 \qquad 6.84 \times 0.04 \qquad 9.11 \times 0.007 \qquad 2.54 \times 0.12 \qquad 8.41 \times 0.24 \qquad 6.52 \times 0.033$$

4.
$$\$0.75 \times 0.8 \qquad \$0.90 \times 0.6 \qquad \$5.00 \times 0.05 \qquad \$8.80 \times 0.75 \qquad \$10.00 \times 0.15 \qquad \$25.00 \times 0.25$$

▶ **Multiply. Round each answer to the hundredths place.**

5. $0.61 \times 0.4 = \qquad 5.6 \times 0.06 = \qquad 2.9 \times 0.007 = \qquad 12 \times 0.009 =$

Answers begin on page 31.

■ MULTIPLYING BY 10, 100, OR 1,000

To *multiply* a number by 10, 100, or 1,000, use one of these shortcuts. You may need to add one or more place-holding zeros before writing the decimal point in the product.

To multiply by 10, move the decimal point **one place** to the right.	To multiply by 100, move the decimal point **two places** to the right.	To multiply by 1,000, move the decimal point **three places** to the right.
$0.7 \times 10 = 07.$ $= 7$	$0.38 \times 100 = 038.$ $= 38$	$0.625 \times 1,000 = 0625.$ $= 625$
$9.4 \times 10 = 94.$ $= 94$	added 0 $17.3 \times 100 = 1730.$ $= 1,730$	added 0s $46.2 \times 1,000 = 46200.$ $= 46,200$

▶ Multiply.

1. $0.8 \times 10 =$ $0.42 \times 10 =$ $10 \times 5.6 =$

2. $0.47 \times 100 =$ $100 \times 3.9 =$ $17.1 \times 100 =$

3. $0.9 \times 1,000 =$ $8.25 \times 1,000 =$ $1,000 \times 31.9 =$

▶ Mixed Multiplication Practice

4.
$\begin{array}{r} 0.9 \\ \times 0.4 \end{array}$
$\begin{array}{r} 2.4 \\ \times 0.7 \end{array}$
$\begin{array}{r} 0.9 \\ \times 7 \end{array}$
$\begin{array}{r} 8.5 \\ \times 6 \end{array}$
$\begin{array}{r} 13.8 \\ \times 10 \end{array}$
$\begin{array}{r} 6.7 \\ \times 100 \end{array}$

5.
$\begin{array}{r} 0.005 \\ \times 0.2 \end{array}$
$\begin{array}{r} 7.7 \\ \times 0.03 \end{array}$
$\begin{array}{r} \$5.00 \\ \times 10 \end{array}$
$\begin{array}{r} \$7.50 \\ \times 0.06 \end{array}$
$\begin{array}{r} 1,000 \\ \times 0.35 \end{array}$
$\begin{array}{r} 7.89 \\ \times 100 \end{array}$

6.
$\begin{array}{r} 12.5 \\ \times 2.3 \end{array}$
$\begin{array}{r} 20 \\ \times 0.05 \end{array}$
$\begin{array}{r} 0.006 \\ \times 4 \end{array}$
$\begin{array}{r} 14.8 \\ \times 0.47 \end{array}$
$\begin{array}{r} \$25.00 \\ \times 0.56 \end{array}$
$\begin{array}{r} \$8.35 \\ \times 100 \end{array}$

7.
$\begin{array}{r} 100 \\ \times 3.7 \end{array}$
$\begin{array}{r} 0.72 \\ \times 8.01 \end{array}$
$\begin{array}{r} 1,000 \\ \times 0.59 \end{array}$
$\begin{array}{r} \$43.99 \\ \times 0.08 \end{array}$
$\begin{array}{r} 0.03 \\ \times 0.14 \end{array}$
$\begin{array}{r} 66.5 \\ \times 1.6 \end{array}$

Answers begin on page 31.

■ APPLYING MULTIPLICATION SKILLS

▶ **Use multiplication to solve each problem.**

1. If round steak is on sale for $2.39 per pound, what would a 6.5-pound package of round steak cost? Round your answer to the nearest cent.

2. A shipping box contains 48 cans of corn. If each can of corn weighs 1.5 pounds, what would be the total weight of the corn in the packed box?

3. Gloria's car gets 18 miles per gallon of gas. Knowing this, determine how far Gloria can travel on a full tank of 16.25 gallons. Round your answer to the nearest mile.

4. Hank placed 3 washers side by side to use as a spacer. What is the total width of Hank's spacer if each washer has a width of 0.0625 inch?

5. For each hour of overtime, Emma earns $8.50 per hour. How much overtime pay will Emma earn on a day in which she works 4.5 overtime hours?

6. In metric measurement, small amounts of weight are measured in grams. It takes 28.4 grams to equal one ounce. Use this information to determine the weight in grams of a parcel that weighs 12 ounces.

7. How much would a customer be charged for a 1.53-pound package of sliced turkey that cost $3.98 per pound?

8. There are 2.54 centimeters in 1 inch. Knowing this, find the width in centimeters of a steel beam that is 6.75 inches wide. Round your answer to the nearest hundredth of a centimeter.

9. Oak countertop molding is on sale for $1.29 per foot. During this sale, what will be charged for a piece of oak molding that is 10 feet long?

10. For his office supply business, Frederick ordered 100 reams of typing paper. If he paid $8.99 per ream, what total amount did Frederick pay for this paper?

11. It costs the AMZ company $6.78 to make its simplest calculator model. How much does it cost AMZ to make 1,000 of these calculators?

12. The Jacksons pay $0.0984 per kilowatt-hour of electricity. Which of the following is the *best estimate* of the Jacksons' electric bill during a month when they use 2,075 kilowatt-hours of electric power?

(1) $0.001 × 2,000
(2) $0.009 × 2,000
(3) $0.01 × 2,000
(4) $0.09 × 2,000
(5) $0.10 × 2,000

Answers begin on page 31.

■ DIVIDING DECIMALS BY WHOLE NUMBERS

To divide a decimal by a whole number:
• Place a decimal point in the quotient directly above its position in the **dividend**.
• Divide, using one or more zeros as placeholders in the dividend as needed.

Example 1. Divide 28.5 by 5.

Place a decimal point in the quotient.
Divide as you do with whole numbers.

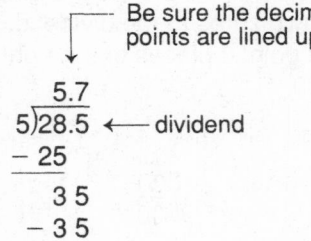

```
          5.7
      5)28.5  ←— dividend
       − 25
          3 5
         − 3 5
```

Example 2. Divide 7 into 0.154.

Place a decimal point in the quotient.
Since you can't divide 7 into 1, write
a zero above the 1. Now divide 7 into 15.

Write a leading zero in the answer.

```
          ↓
         0.022
     7)0.154
      − 14
          14
        − 14
```

▶ **Divide.**

1. 4)4.48   5)0.655   3)18.6   12)0.528   14)4.284

2. 6)0.276   8)0.424   6)0.882   13)1.105   27)0.0567

3. 4)$10.16   3)$7.29   5)$12.75   4)$0.20   8)$0.32

▶ **In the following problems, use a zero as a placeholder and then continue to divide.**

4. 3)0.087   12)0.108   9)0.486   7)0.0063   4)0.0504

5. 8)0.4   4)0.2   28)1.4   80)4.8   60)1.2

Answers begin on page 31.

22

■ DIVIDING DECIMALS BY DECIMALS

To divide a decimal by a decimal:
* Change the divisor to a whole number by moving its decimal point to the far right.
* Move the decimal point in the dividend an equal number of places to the right, adding place-holding zeros if needed.
* Divide.

Example 1. $0.4\overline{)6.52}$

Move the decimal point in the divisor and dividend 1 place to the right. Divide 4 into 65.2.

$0\,4.\overline{)6\,5.\,2}$

No added zeros.

$$\begin{array}{r} 16.3 \\ 4\overline{)65.2} \\ -4 \\ \hline 25 \\ -24 \\ \hline 1\,2 \\ -1\,2 \\ \hline \end{array}$$

Example 2. $0.005\overline{)2.5}$

Move the decimal point in the divisor 3 places to the right. Add two zeros to the dividend, and move its decimal point 3 places to the right. Divide 5 into 2500.

Two added zeros.

$0\,005.\overline{)2\,500.}$

$$\begin{array}{r} 500 \\ 5\overline{)2500} \\ -25 \\ \hline 000 \\ 00 \\ \hline \end{array}$$

▶ **Divide.**

1. $0.2\overline{)0.38}$ $0.6\overline{)0.42}$ $0.3\overline{)5.4}$ $1.2\overline{)2.64}$ $5.6\overline{)8.96}$

2. $0.03\overline{)1.5}$ $0.014\overline{)30.8}$ $4.7\overline{)14.899}$ $2.5\overline{)26.5}$ $3.6\overline{)0.01944}$

▶ **To divide a whole number by a decimal, write a decimal point *to the right* of the whole number. Then use place-holding zeros and move the decimal point the appropriate number of places.**

3. $0.07\overline{)21}$ $0.03\overline{)18}$ $0.05\overline{)10}$ $0.8\overline{)16}$ $0.4\overline{)24}$

4. $2.5\overline{)50}$ $1.2\overline{)48}$ $4.2\overline{)840}$ $1.5\overline{)315}$ $2.3\overline{)483}$

Answers begin on page 32.

■ DIVIDING BY 10, 100, OR 1,000

To *divide* a number by 10, 100, or 1,000, use one of these shortcuts. You may need to add one or more place-holding zeros before writing the decimal point in the quotient.

To divide by 10, move the decimal point **one place** to the left.	To divide by 100, move the decimal point **two places** to the left.	To divide by 1,000, move the decimal point **three places** to the left.
0.5 ÷ 10 = .0 5 = 0.05	30.8 ÷ 100 = .30 8 = 0.308	147 ÷ 1,000 = .147 = 0.147
3.7 ÷ 10 = .3 7 = 0.37	⌐ added zero 7.4 ÷ 100 = .07 4 = 0.074	⌐⌐ added zeros 9.2 ÷ 1,000 = .009 2 = 0.0092

▶ **Divide.**

1. 0.8 ÷ 10 =   4.5 ÷ 10 =   23.7 ÷ 10 =

2. 31.4 ÷ 100 =   9.6 ÷ 100 =   0.4 ÷ 100 =

3. 253 ÷ 1,000 =   39 ÷ 1,000 =   6.5 ÷ 1,000 =

▶ **Mixed Division Practice**

4. 2)0.842   6)0.612   4)$43.36   5)46.65   14)32.76

5. 7)0.427   27)0.0567   4)3   5)6   16)7

6. 0.02)1.6   3.1)0.093   1.2)27.84   0.05)1.5   0.035)0.497

Answers begin on page 32.

■ APPLYING DIVISION SKILLS

▶ **Use division to solve each problem.**

1. Jenny and two friends agreed to equally share the cost of renting a canoe for Saturday. If the total cost is $33.45, what is Jenny's share?

2. A 1.4-ounce bottle of French perfume is on sale for $34.93. At this rate, what would a customer be charged per ounce of this perfume?

3. Myra paid $12.84 for a 7-pound turkey. To the nearest cent, how much did Myra pay per pound?

4. The width of a human hair is measured to be 0.012 inch. How many similar hairs could be placed side by side in a 1.5-inch section of scalp?

5. While traveling through Oregon, Daryl drove 404.25 miles on 16.5 gallons of gas. Knowing this, determine his car's mileage—the number of miles his car can go on one gallon.

6. Sherm's construction crew can repair 0.38 mile of road per day. How long will it take Sherm's crew to repair an 8.4-mile section of road working at this rate? Choose the best answer from the choices below.

    (1) a little less than 21 days
    (2) about 21.5 days
    (3) a little more than 22 days
    (4) about 23 days
    (5) a little more than 24 days

7. The price of a 3.5-pound package of chicken is $4.69. At this rate, what is the cost of chicken per pound?

8. A jeweler has 1.65 ounces of gold with which to make 7 rings. To the nearest hundredth of an ounce, how much gold should he use for each ring?

9. Lucy has a red ribbon that is 135 inches long. She wants to cut this ribbon into 10 equal pieces. How long should she cut each piece?

10. A box containing 100 small dictionaries weighs 54.68 kilograms, not counting the weight of the box itself. What is the weight of each dictionary to the nearest thousandth of a kilogram?

11. Manuel bought 1,000 pens to give away to customers at his auto supply store. If he paid $396 for the pens, how much did each pen cost Manuel? Round your answer to the nearest cent.

12. Erik frequently changes the oil in his customers' cars. At present, Erik has a container of 322.5 liters of new oil. If an oil change averages 4.19 liters, what division **best estimates** how many oil changes Erik has enough oil for?

    (1) 320 ÷ 4
    (2) 300 ÷ 4
    (3) 330 ÷ 4
    (4) 320 ÷ 5
    (5) 330 ÷ 5

Answers begin on page 32.

■ CHANGING A FRACTION TO A DECIMAL

To change a fraction into a decimal, divide the denominator into the numerator.
• Write a decimal point to the right of the dividend (**numerator**) and add place-holding zeros as needed.
• Divide until there is no remainder or until you have all the decimal places you want.

Example 1. Change $\frac{2}{5}$ to a decimal.

Divide 5 into 2. Add one place-holding zero.
Place a decimal point in the quotient.
Divide, and write a leading zero in the answer.

$$
\begin{array}{r}
0.4 \\
5\overline{)2.0} \\
-2\,0 \\
\end{array}
$$

Example 2. Change $\frac{3}{4}$ to a decimal.

Divide 4 into 3. Add two place-holding zeros.
Place a decimal point in the quotient.
Divide, and write a leading zero in the answer.

$$
\begin{array}{r}
0.75 \\
4\overline{)3.00} \\
-2\,8 \\
\hline
20 \\
-20 \\
\end{array}
$$

▶ **Divide. Add enough place-holding zeros so that each answer ends without a remainder.**

1. $5\overline{)3}$ $\qquad$ $4\overline{)1}$ $\qquad$ $12\overline{)9}$ $\qquad$ $8\overline{)7}$ $\qquad$ $16\overline{)3}$

▶ **Change each fraction to a decimal with no remainder. Before dividing, reduce proper fractions to lowest terms.**

2. $\frac{2}{4} =$ $\qquad$ $\frac{6}{8} =$ $\qquad$ $\frac{14}{20} =$ $\qquad$ $\frac{5}{8} =$ $\qquad$ $\frac{7}{16} =$

A **repeating decimal** occurs when you divide two numbers and continue to get a remainder with a repeating pattern. For example: $\frac{1}{3} = 0.3333333\ldots$ Each fraction below has a repeating decimal equivalent. Divide, and round each answer to the hundredths place.

> Two ways to represent a repeating decimal:
> $\frac{1}{3} = 0.3\ldots$ *or* $0.\overline{3}$

3. $\frac{1}{3} \approx$ $\qquad$ $\frac{2}{3} \approx$ $\qquad$ $\frac{1}{6} \approx$ $\qquad$ $\frac{5}{9} \approx$ $\qquad$ $\frac{3}{11} \approx$

Answers begin on page 32.

26

■ WORKING WITH FRACTIONS AND DECIMALS

Comparing Decimal Fractions with Proper Fractions

To compare decimal fractions with **proper fractions**, change the proper fraction to a decimal. With both numbers written as decimals, follow the rules for comparing decimal fractions given on page 6.

Example 1. How much larger is $\frac{3}{4}$ than 0.68?

Step 1. Write $\frac{3}{4}$ as a decimal fraction.

Dividing 4 into 3, we get $\frac{3}{4} = 0.75$.

Step 2. Subtract 0.68 from 0.75.

$$\begin{array}{r} 0.75 \\ -\ 0.68 \\ \hline 0.07 \end{array}$$

$\frac{3}{4}$ is larger than 0.68 by 0.07.

Example 2. By how much do 1.648 and $1\frac{5}{8}$ differ?

Step 1. Write $\frac{5}{8}$ as a decimal fraction.

Dividing 8 into 5, we get $\frac{5}{8} = 0.625$.

Step 2. Subtract 1.625 $\left(1\frac{5}{8}\right)$ from 1.648.

$$\begin{array}{r} 1.648 \\ -\ 1.625 \\ \hline 0.023 \end{array}$$

1.648 and $1\frac{5}{8}$ differ by 0.023.

▶ **For rows 1 and 2, circle the larger value in each pair.**

1. $\frac{7}{8}$ or 0.859　　　0.249 or $\frac{1}{4}$　　　0.131 or $\frac{1}{8}$　　　$\frac{4}{5}$ or 0.82

To make comparisons in row 2, your first step is to round each fraction to three decimal places.

2. $\frac{1}{3}$ or 0.336　　　$\frac{5}{6}$ or 0.81　　　$\frac{8}{11}$ or 0.7　　　$\frac{1}{9}$ or 0.12

3. Which is the larger amount: $0.43 or $\frac{7}{16}$ of a dollar?

4. Knowing that 1 yard = 36 inches, circle the longer distance. 13 inches or $\frac{1}{3}$ yard.

5. Which is the smaller area: $\frac{7}{8}$ square foot or 0.8 square foot?

6. Knowing that 1 pound = 16 ounces, circle the heavier amount. 7 ounces or 0.45 pound.

Comparison Word Problems

To solve a comparison word problem, rewrite each fraction as a decimal and then compare decimals as needed. In problems involving money, round each answer to the nearest cent.

▶ **Solve each comparison problem below.**

7. a) Will a pipe that is $4\frac{5}{8}$ inches wide fit through a hole that is 4.65 inches across?

   b) What is the difference in width between the pipe and the hole?

8. While shopping at Shopper's Mart, Rhonda bought $1\frac{3}{4}$ pounds of apples for $1.50 per pound. How much did Rhonda pay for these apples?

9. On his afternoon jog, Sterling ran 1.5 miles in $12\frac{3}{4}$ minutes. On the average, how long did it take Sterling to run each mile?

10. Two-inch-wide molding costs $0.42 per foot of length. How much does a piece of molding that is 3 feet, 9 inches long cost? (Hint: As your first step, change 9 inches to a fraction of a foot.)

11. At Shelley's Fabrics, Ellen bought $4\frac{1}{4}$ yards of fabric for $3.25 per yard. How much was Ellen charged for this fabric?

12. According to the blueprint, Alan is to drill a drainage hole no larger than 0.6 inch in diameter. He has the three drill-bit sizes shown below to choose from. Which bit will give him the largest permissible hole?

Bit no.	Diameter
1	$\frac{1}{2}$ inch
2	$\frac{9}{16}$ inch
3	$\frac{5}{8}$ inch

13. a) Express each weight below as a decimal rounded to two decimal places.

   A: $\frac{2}{3}$ ounce _____

   B: 0.587 ounce _____

   C: $\frac{7}{12}$ ounce _____

   D: $\frac{3}{5}$ ounce _____

   b) Write these weights in order. Write the letter of the *lightest weight at the left* and the heaviest at the right.

   _____ _____ _____ _____

14. On Saturday, Warren worked $3\frac{3}{4}$ hours of overtime and was paid $9.60 per hour. How much did Warren earn for these overtime hours?

Answers begin on page 32.

◪ DECIMAL SKILLS REVIEW

▶ **Round each decimal to the place indicated.**

1. $0.78 \approx$ $0.325 \approx$ $14.752 \approx$ $4.6567 \approx$
     tenths   hundredths   hundredths   thousandths

▶ **Add.**

2.
0.5	0.8	1.4	1.35	2.37	$12.65
+ 0.3	+ 0.7	+ 0.9	+ 0.62	+ 1.25	+ 4.00

3.
2.5	1.25	4.6	5	4.82	$5.00
+ 1.75	+ 0.8	+ 3	2.4	3.5	3.78
			+ 0.95	+ 2	+ 1.49

▶ **Subtract.**

4.
0.6	8.7	1.37	21.49	$5.35	$15.00
− 0.2	− 5.9	− 0.42	− 9.75	− 2.80	− 6.49

5.
0.84	0.325	5.4	3	$6.78	$20.00
− 0.7	− 0.1	− 3.25	− 0.6	− 2.50	− 7.56

▶ **Multiply.**

6.
0.9	6.4	9.23	$0.90	$7.50	8.24
× 6	× 7	× 4	× 6	× 12	× 15

7.
7.5	$3.45	0.006	84.5	100	1,000
× 6.5	× 0.4	× 0.8	× 10	× 0.75	× 5.7

▶ **Divide.**

8. $6\overline{)12.54}$ $5\overline{)\$14.85}$ $0.03\overline{)0.051}$ $0.6\overline{)2.88}$ $1.9\overline{)4.94}$

9. $0.04\overline{)0.64}$ $13\overline{)0.026}$ $0.7 \div 10 =$ $14.9 \div 100 =$ $2.7 \div 1,000 =$

▶ **Solve each problem.**

10. At a picnic, two tables were placed end to end. The first table was 2.8 meters long, and the second table was 2.65 meters long. What was the combined length of the two tables?

11. Annie lives 0.5 mile closer to work than her friend Georgia does. If Georgia lives 4.75 miles from work, how far from work does Annie live?

12. Carolyn is mailing three boxes. The first weighs 1.625 pounds, the second weighs 0.75 pound, and the third weighs 1.25 pounds. What is the total weight of Carolyn's three boxes?

13. A gallon of water weighs about 8.36 pounds. To the nearest pound, what is the weight of water in a full waterbed if the bed holds 64 gallons of water?

14. How much change should Allie receive from a $17.84 purchase if she pays the clerk with a check for $30.00?

15. One state's sales tax is 0.065 cent for each dollar of purchase. In this state, how much sales tax to the nearest cent will be charged on a purchase of $37.98?

16. The area of the United States is 3.62 million square miles, while the area of Mexico is 0.76 million square miles. By how many million square miles is the United States larger than Mexico?

17. Noreen worked 28.6 hours last week. At $6.50 per hour, how much did Noreen earn last week?

18. Juanita ran the 100-meter dash in 12.65 seconds. Her sister Mandy ran the same race in 13.09 seconds. What is the difference between the race times of the two sisters?

19. One mile equals approximately 1.61 kilometers. To the nearest tenth of a kilometer, how many kilometers is equivalent to a distance of 35 miles?

20. On the way to Chicago, the Malone family drove 312.8 miles on 12.5 gallons of gas. What was their car's average mileage (miles per gallon) for this distance? Round your answer to the nearest mile per gallon.

21. At Friendly Market, Ellie bought a 3.8-pound roast for $10.60. To the nearest cent, what was Ellie charged per pound for this roast?

Answers begin on page 32.

■ ANSWER KEY

Page 3
1. 0.2
2. 0.06
3. 0.009
4. 0.5
5. 0.07
6. 0.003

Page 4
1. four-tenths, eight-tenths
2. seven-hundredths, fourteen-hundredths
3. thirty-hundredths, fifty-hundredths
4. sixty-two thousandths, twenty-three thousandths
5. seventy-thousandths, ninety-thousandths
6. one hundred forty-six thousandths, two hundred seventy-eight thousandths
7. three hundred fifty-thousandths, six hundred thousandths
8. 0.04 and 0.040
9. 0.120 and 0.12
10. 0.70 and 0.7

Page 5
1. 0.5, 0.7, 0.9
2. 0.08, 0.09, 0.12
3. 0.10, 0.40, 0.90
4. 0.005, 0.023, 0.080
5. 0.275, 0.750, 0.500
6. 0.35, 0.190
7. 0.8, 0.60
8. 0.340, 0.502
9. 0.04, 0.207
10. 0.38
11. 0.6
12. 0.91
13. 0.394
14. 0.45

Page 6
A. 1. <, >, <
   2. >, <, <
   3. <, =, >
B. 1st: Sunripe at $0.089
   2nd: Emerald at $0.09
   3rd: OJ Plus as $0.094
   4th: Valley D at $0.102
C. A: Drawer 3
   B: Drawer 1
   C: Drawer 3
   D: Drawer 2
   E: Drawer 1

Page 7
1. 5.3, 15.32
2. 6.019, 3.08
3. 74.225
4. 126.746
5. five and seven-tenths, six and fifteen-hundredths
6. sixty-four and three hundred seventy-five thousandths
7. one hundred twenty-eight and ninety-two hundredths
8. Race 1: 1st: Alan at 58.234 sec.
          2nd: Sal at 58.79 sec.
          3rd: Don at 59.06 sec.
          4th: Vic at 59.084 sec.
   Race 2: 1st: Jani at 43.7 sec.
          2nd: Jolene at 43.78 sec.
          3rd: Stacey at 44.65 sec.
          4th: Ellie at 44.732 sec.

Pages 8–9
1. 6 meters, $8, 10 centimeters
2. $17, 8 ounces, 26 miles per gallon
3. (3) 14 + 10
4. (3) 35 − 13
5. (1) 14 × 9
6. (1) 34 ÷ 6

Answers to 7 and 8 may vary. Any reasonable answer is acceptable. Here are some sample answers:
7. $14, 3, $170
8. 13, 5, 3
9. 16.955
10. 11.685
11. 93.4825
12. 405.82
13. 7.63
14. 31.87
15. (3) 10.975 kg
16. (2) $285.80
17. (5) $24.49
18. (1) 20.2

Page 11
1. 0.4, 0.3, 0.5, 0.1, 0.8
2. 3.3, 1.9, 2.8, 3.1, 8.0
3. 0.24, 0.81, 0.35, 0.89, 0.25
4. 2.83, 1.05, 6.50, 7.81, 21.88
5. 0.474, 0.973, 1.947, 3.489, 17.338
6. 0.308, 0.640, 4.251, 6.009, 12.105
7. (a) $7.41
   (b) $15.55
   (c) $15.13
   (d) $11.25
   (e) $6.89
   (f) $12.79
   (g) (e) is incorrect. By estimating (2 × $5) you can see that the correct answer is a little more than $10.

Page 12
1. 1.5, 1.5, 0.85, 2.61, 1.331, 3.003
2. 9.1, 27.5, 36.37, 56.32, 13.069, 23.208
3. $1.10, $1.46, $3.07, $5.20, $10.78, $34.64
4. 3.7, 17.2, $18.90, $71.41, 8.604, 25.545
5. 1.31 ≈ 1.3, 4.37 ≈ 4.4, 12.47 ≈ 12.5, 32.99 ≈ 33.0
6. $2.40, $78.50, $12.94

Page 13
1. 0.95, 1.65, 1.07, 1.735, 0.625, 1.365
2. 7.2, 9.1, 12.5, 12.8, 8.33, 16.667
3. 8.56, 10.74, 14.125, 6.225, 12.22, 33.14
4. 6.55, 18.75, 16.07, 32.05, 8.15, 28.483
5. 14.475 ≈ 14.48, 13.375 ≈ 13.38, 14.509 ≈ 14.51
6. $4.75, $14.25, $27.86

Pages 14–15
1. 0.4, 0.5, 0.62, 0.38, .218, 0.495
2. 3.1, 5.2, 1.9, 4.22, 7.85, 7.75
3. 6.58, 1.57, 2.25, 8.951, 4.618, 7.25
4. $0.62, $0.42, $0.89, $1.89, $4.02, $8.11
5. 0.7, 6.8, 6.6, 7.78
6. $2.85, $12.23, $19.74
7. 0.15, 0.54, 0.125, 0.345, 0.354, 0.325
8. 0.45, 0.18, 0.13, 0.275, 0.575, 0.555
9. 5.8, 1.5, 2.25, 8.375, 2.265, 14.75
10. 0.25, 0.275, 0.15, 0.475
11. 0.97, 1.17, 4.2, 4.625
12. $4.25, $3.57, $11.21

Page 16
1. $2.57 for two gallons
2. .309 batting average
3. 1.727 meters
4. 6.55 ≈ 6.6 miles
5. $24.96
6. 0.015625 ≈ 0.016 inch
7. 0.125 inch

8. 4.4°F
9. $3.51 change
10. 32 hours
11. $0.03 less per kilowatt-hour
12. (3) 6 + 4

Page 17
1. 4.0 = 4, 3.6, 4.2, 3.2, 6.3, 5.6
2. 19.5, 50.4, 37.6, 52.2, 46.8, 38.0 = 38
3. 27.36, 15.08, 112.70 = 112.7, 114.48, 111.375, 183.477
4. $4.00, $4.50, $4.32, $46.75, $238.00, $210.00
5. 7.28 ≈ 7.3, 58.24 ≈ 58.2, 39.12 ≈ 39.1, 129.375 ≈ 129.4

Page 18
1. 0.24, 0.14, 0.40 = 0.4, 0.054, 0.01, 0.0036
2. 3.00 = 3, 5.58, 4.13, 0.544, 0.270 = 0.27, 0.0595
3. 0.171, 0.2736, 0.06377, 0.3048, 2.0184, 0.21516
4. $0.60, $0.54, $0.25, $6.60, $1.50, $6.25
5. 0.244 ≈ 0.24, 0.336 ≈ 0.34, 0.0203 ≈ 0.02, 0.108 ≈ 0.11

Page 19
1. 8, 4.2, 56
2. 47; 390; 1,710
3. 900; 8,250; 31,900
4. 0.36, 1.68, 6.3, 51, 138, 670
5. 0.001, 0.231, $50, $0.45, 350, 789
6. 28.75, 1, 0.024, 6.956, 14, $835
7. 370, 5.7672, 590, 3,5192, 0.0042, 106.4

Page 20
1. $15.54
2. 72 pounds
3. 292.5 ≈ 293 miles
4. 0.1875 inch
5. $38.25
6. 340.8 grams
7. $6.09
8. 17.145 ≈ 17.15 centimeters
9. $12.90
10. $899
11. $6,780
12. (5) $0.10 × 2,000

Page 21
1. 1.12, 0.131, 6.2, 0.044, 0.306
2. 0.046, 0.053, 0.147, 0.085, 0.0021
3. $2.54, $2.43, $2.55, $0.05, $0.04
4. 0.029, 0.009, 0.054, .0009, .0126
5. 0.05, 0.05, 0.05, 0.06, 0.02

Page 22
1. 1.9, 0.7, 18, 2.2, 1.6
2. 50; 2,200; 3.17; 10.6; 0.0054
3. 300, 600, 200, 20, 60
4. 20, 40, 200, 210, 210

Page 23
1. 0.08, 0.45, 2.37
2. 0.314, 0.096, 0.004
3. 0.253, 0.039, 0.0065
4. 0.421, 0.102, $10.84, 9.33, 2.34
5. 0.061, 0.0021, 0.75, 1.2, 0.4375
6. 80, 0.03, 23.2, 30, 14.2

Page 24
1. $11.15
2. $24.95 per ounce
3. $1.83 per pound
4. 125 hairs
5. 24.5 miles per gallon
6. (3) a little more than 22 days
7. $1.34 per pound
8. $0.235 \approx 0.24$ ounce
9. 13.5 inches
10. $0.5468 \approx 0.547$ kilogram
11. $0.396 \approx$ $0.40 each
12. (1) $320 \div 4$

Page 25
1. 0.6, 0.25, 0.75, 0.875, 0.1875
2. .5, 0.75, 0.7, 0.625, 0.4375
3. 0.33, 0.67, 0.17, 0.56, 0.27

Pages 26–27
1. $\frac{7}{8}$(0.875), $\frac{1}{4}$(0.25), 0.131, 0.82
2. 0.336, $\frac{5}{6}$(0.8333 . . .), $\frac{8}{11}$(0.72 . . .), 0.12
3. $\frac{7}{16}$ of a dollar is larger.
4. 13 inches is longer.

5. 0.8 square foot is the smaller area.
6. 0.45 pound is the heavier amount.
7. a) yes
   b) 0.025 in. (4.65 − 4.625)
8. $2.63
9. 8.5 minutes per mile
10. $1.58
11. $13.81
12. Bit 2, $\frac{9}{16}$ inch
13. a) A: 0.67 ounce
    B: 0.59 ounce
    C: 0.58 ounce
    D: 0.6 ounce
    b) C, B, D, A
14. $36.00

Pages 28–29
1. 0.8, 0.33, 14.75, 4.657
2. 0.8, 1.5, 2.3, 1.97, 3.62, $16.65
3. 4.25, 2.05, 7.6, 8.35, 10.32, $10.27
4. 0.4, 2.8, 0.95, 11.74, $2.55, $8.51
5. 0.14, 0.225, 2.15, 2.4, $4.28, $12.44
6. 5.4, 44.8, 36.92, $5.40, $90, 123.6
7. 48.75; $1.38; 0.0048; 845; 75; 5,700
8. 2.09, $2.97, 1.7, 4.8, 2.6
9. 16, 0.002, 0.07, 0.149, 0.0027
10. 5.45 meters
11. 4.25 miles
12. 3.625 pounds
13. 535 pounds
14. $12.16
15. $2.47
16. 2.86 million square miles
17. $185.90
18. 0.44 second
19. 56.4 kilometers
20. 25 miles per gallon
21. $2.79

Available only in packages of 10. To order, use number 3691-5.

ISBN 0-8092-3826-8

9 780809 238262

90000